AF885241

Learn How To Write

Trace and Count Workbook

Numbers

ISBN: 978-1-7372625-2-7
Creator: Marsheka Jones

Number Tracing

Name: ____________________ Date: ____________________

One

1

1 1 1 1 1 1

1 1 1 1 1 1

1 1 1 1 1 1

ONE ONE ONE ONE

ONE ONE ONE ONE

ONE ONE ONE ONE

One

Name:

Date:

Two

2

2 2 2 2 2 2

2 2 2 2 2 2

2 2 2 2 2 2

TWO TWO TWO TWO

TWO TWO TWO TWO

TWO TWO TWO TWO

Two

Name: ____________________ Date: ____________________

Three

3

3 3 3 3 3 3

3 3 3 3 3 3

3 3 3 3 3 3

THREE THREE THREE

THREE THREE THREE

THREE THREE THREE

Three

Name:

Date:

Four

4

FOUR FOUR FOUR

FOUR FOUR FOUR

FOUR FOUR FOUR

Four

Name:

Date:

Five

5

5 5 5 5 5 5

5 5 5 5 5 5

5 5 5 5 5 5

FIVE FIVE FIVE FIVE

FIVE FIVE FIVE FIVE

FIVE FIVE FIVE FIVE

Five

Name:

Date:

Six

6

6 6 6 6 6

6 6 6 6 6

6 6 6 6 6

SIX SIX SIX SIX SIX SIX

SIX SIX SIX SIX SIX SIX

SIX SIX SIX SIX SIX SIX

Six

Name:

Date:

Seven

7

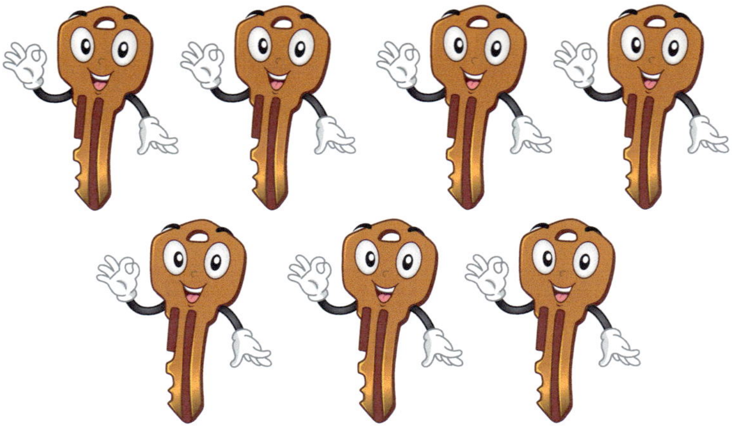

SEVEN SEVEN SEVEN

SEVEN SEVEN SEVEN

SEVEN SEVEN SEVEN

Seven

Name:

Date:

Eight

8

8 8 8 8 8

8 8 8 8 8

8 8 8 8 8

EIGHT EIGHT EIGHT

EIGHT EIGHT EIGHT

EIGHT EIGHT EIGHT

Eight

Name:

Date:

Nine

9

NINE NINE NINE NINE

NINE NINE NINE NINE

NINE NINE NINE NINE

Nine

Name:

Date:

Zero

0

0 0 0 0 0

0 0 0 0 0

0 0 0 0 0

ZERO ZERO ZERO ZERO
ZERO ZERO ZERO ZERO
ZERO ZERO ZERO ZERO

Zero 0

More Tracing Practice

Name: ____________________ Date: ____________________

TRACE THE NUMBERS

1	1 1 1 1 1 1
2	2 2 2 2 2
3	3 3 3 3 3
4	4 4 4 4 4
5	5 5 5 5 5

Name: ______________________ Date: ______________________

TRACE THE NUMBERS

6	6 6 6 6 6
7	7 7 7 7 7
8	8 8 8 8 8
9	9 9 9 9 9
0	0 0 0 0 0

Name: ______________________________

NUMBER TRAIN

Practice writing numbers by tracing the number in each car of this train.

Tracing Numbers 1-100

TRACING NUMBERS 1-20

1	2	3	4
5	6	7	8
9	10	11	12
13	14	15	16
17	18	19	20

TRACING NUMBERS 21-40

21	22	23	24
25	26	27	28
29	30	31	32
33	34	35	36
37	38	39	40

TRACING 41-60 NUMBERS

41	42	43	44
45	46	47	48
49	50	51	52
53	54	55	56
57	58	59	60

TRACING NUMBERS 61-80

61	62	63	64
65	66	67	68
69	70	71	72
73	74	75	76
77	78	79	80

TRACING NUMBERS 81-100

81	82	83	84
85	86	87	88
89	90	91	92
93	94	95	96
97	98	99	100

Identifying Numbers

Name: ______________________ Date: ______________________

FIND THE NUMBER.

DIRECTIONS: TRACE THE NUMBER THEN COLOR THE CIRCLES THAT HAVE THE NUMBER YOU TRACED.

1 ONE

2 1

1 3

6 1 1

7

5 1

Name: ____________________ Date: ____________________

FIND THE NUMBER.

DIRECTIONS: TRACE THE NUMBER THEN COLOR THE CIRCLES THAT HAVE THE NUMBER YOU TRACED.

2 two

2 2

1 3

6 2

1

2

2 1

Name: ____________________ Date: ____________________

FIND THE NUMBER.

DIRECTIONS: TRACE THE NUMBER THEN COLOR THE CIRCLES THAT HAVE THE NUMBER YOU TRACED.

3 THREE

2 3 2 3 6 3 1 2 3 3 3

Name:

Date:

FIND THE NUMBER.

DIRECTIONS: TRACE THE NUMBER THEN COLOR THE CIRCLES THAT HAVE THE NUMBER YOU TRACED.

4 FOUR

4

4

9

1

6

4

4

4

7

3

Name: ______________________ Date: ______________________

FIND THE NUMBER.

DIRECTIONS: TRACE THE NUMBER THEN COLOR THE CIRCLES THAT HAVE THE NUMBER YOU TRACED.

5 FIVE

5 5

9 1

6 5

5

5

7

3

Name: ____________________ Date: ____________________

FIND THE NUMBER.

DIRECTIONS: TRACE THE NUMBER THEN COLOR THE CIRCLES THAT HAVE THE NUMBER YOU TRACED.

6 SIX

4 6

6 1

6 4

8

3

6 0

Name: ______________________ Date: ______________________

FIND THE NUMBER.

DIRECTIONS: TRACE THE NUMBER THEN COLOR THE CIRCLES THAT HAVE THE NUMBER YOU TRACED.

7 SEVEN

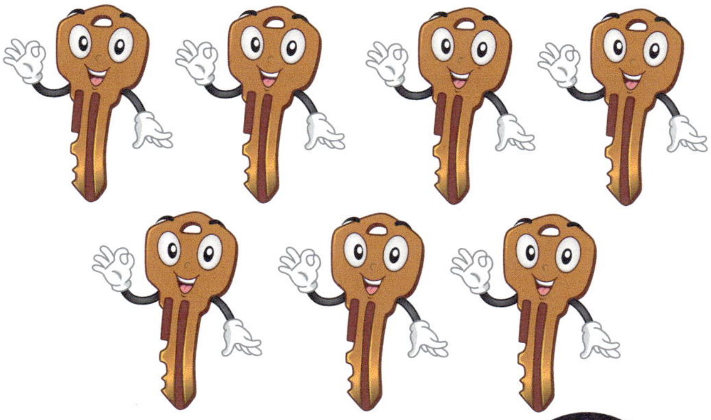

7 9 1 7 6 4 7 7 7 7

Name: ______________________ Date: ______________________

FIND THE NUMBER.

DIRECTIONS: TRACE THE NUMBER THEN COLOR THE CIRCLES THAT HAVE THE NUMBER YOU TRACED.

8 EIGHT

8 8 9 1 6 8 8 8 7 3

Name:

Date:

FIND THE NUMBER.

DIRECTIONS: TRACE THE NUMBER THEN COLOR THE CIRCLES THAT HAVE THE NUMBER YOU TRACED.

9 NINE

4

3

9

2

6

9

1

9

7

9

Name:

Date:

FIND THE NUMBER.

DIRECTIONS: TRACE THE NUMBER THEN COLOR THE CIRCLES THAT HAVE THE NUMBER YOU TRACED.

0 ZERO

0

0 9 1 0 6 0 0 0 7 3

Congratulations

This certificate is awarded to

____

For successfully learning how to
write and identify numbers

Marsheka Jones

Marsheka Jones, Creator

____

Date

Made in the USA
Las Vegas, NV
07 November 2026

57672508R00029